The King Is So Happy Today

# 大王今天很高兴

东子 绘·著

舒友文 译

中国青年出版社

## 东子简历

# WANG Dong

| | |
|---|---|
| 2014 | Book 《Da Li Little Things》By China Youth Publishing House |
| 2006 | Book 《Love Letter to the God》 By Hu Nan Art Book Publishing House |
| 2005 | Book 《Kiss Under the Starry Night of Da Li》 By Guangxi People's Publishing House which is Simplified Chinese version |
| | Book 《Kiss Under the Starry Night of Da Li》 By Taiwan Linking Publishing House which is Complicated Chinese version |
| 2004 | Book 《Cigarette Box》 By China Youth Publishing House |
| 2001 | Art Work "Blue China Chinese Character Make Poem" Win Awards at Italian International Faenza Pottery Exhibition |
| 2000 | Book 《French Lady Who Making Pottery》 By Hu Nan Art Book Publishing House |
| | Open Art Shop Zen Cat at Beijing Shisha Hai Lake |
| 1994 | Win French National Art Scholarship and Study Art in France and travel extensively in Europe |
| 1988 | Graduated from Beijing Fine Art University China,major in Sculpture |
| 1967 | Born in Beijing China |

Email    dongzi707@163.com

# Table of Contents

# 目录

感恩

一切都是老天爷的赐福

*Be Grateful*

*Everything was blessed by the Heavens.*

昨晚梦见巨大的古道

昨晚梦见巨大的古莲

*I dreamed huge ancient lotus last night.*

穿越前世今生

让我们一起在浪里游一段路程

*Through the Past and Present*

*Let us swim our journey in the waves.*

听我喜欢的歌与我一同飞行

*Listen to my favorite songs and fly with me.*

给我讲个故事

我来找你

*Tell me some stories*

*I will come over to you.*

此刻已是蝴蝶梦

*In the butterfly dream.*

学习松的忍耐

才是成长的开始吧

*Learning restraint character from pine trees*

*That is the start of grow-up?*

心中有期待

真好

*How nice it is to have expectation and hope in heart.*

心被爱打开

最远的地方可能有最近的路

*Heart opens up by love.*

*The furthermost place might have the nearest*

*road.*

生活如同一个空的礼物盒子
无论经历什么都不会带走一丝一毫
我们将用生命完成对空的赞美并感恩

*Life like an empty gift box*

*Nothing will be taken away no matter what we*

*had experienced and will experience*

*We will use our life to be grateful and give*

*compliment to the emptiness.*

人类被时间捆绑了

*Human being has been kidnapped by TIME.*

人生是一次非计划的临时行为

*Life is an unplanned provisional behavior.*

我们将穷其一生的劳作来埋藏创造出来的自我

*We will bury the created self with our full-life long work.*

学会控制才能重返自由

*Learning control would help you return to freedom.*

快乐是对自己最大的慈善

*Finding happiness is the largest philanthropic to yourself.*

## 自由是需要技巧的

*Freedom needs skill.*

带着爱说出真话是安全的

*It is secure to speak honest words with love.*

心中有爱通向未来

*With love in the heart*
*it leads us to the future.*

教育的目的是为了告诉我们
不要被人生的假象所摧毁

*The purpose of education is to tell us*

*We cannot be destroyed by the illusion of life.*

我活着是替你在活

*I am alive now, is to be alive on your behalf.*

古今感应

凭今感应

*At this very moment*

*I start feel.*

別急

*Take it easy.*

44

安静地坐在阳光里

*Sit quietly in the sunshine.*

你有没有发现十年很快就过去了

而每一天却很慢

*Have you noticed time fly so fast even if it is a*

*decade,*

*However it is so slow to spend every and each*

*single day.*

这个世界
就是这个世界的
不是谁的
*This world is this world*
*It did not belong to anyone.*

天好，被子全晒在太阳光里

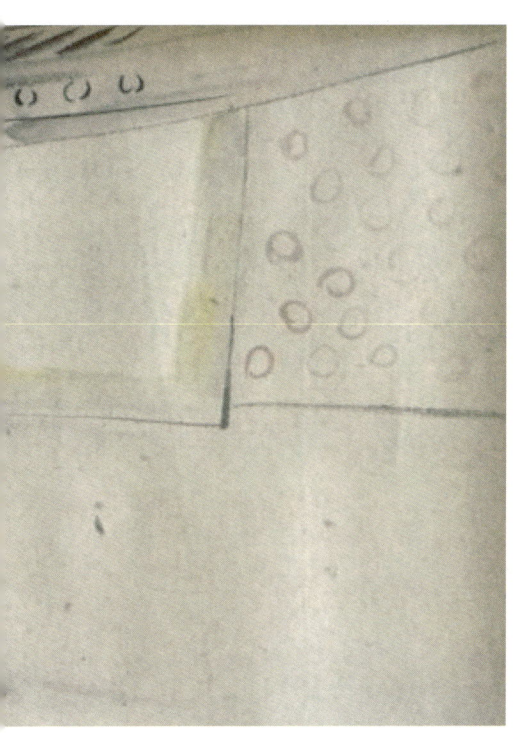

心情和被子一样
得经常拿出来晒

*Mood and quilt are the*
*same way,*
*They would needed to*
*be frequently drying out*
*under the Sun.*

所有一切的精彩表演都是为了获得爱的能量

*All the excellent performance*

*In order to get the energy of love.*

平静在哪呢

*Where is tranquility.*

人生不易

因另有其意

*Life is never easy*

*because of its deep meaning.*

笔墨入心夺其香

*Ink and brushes wins their fragrance*
*when the art work catches our heart.*

爱的越深

离得越远

*The deeper one loves*

*The further the distance.*

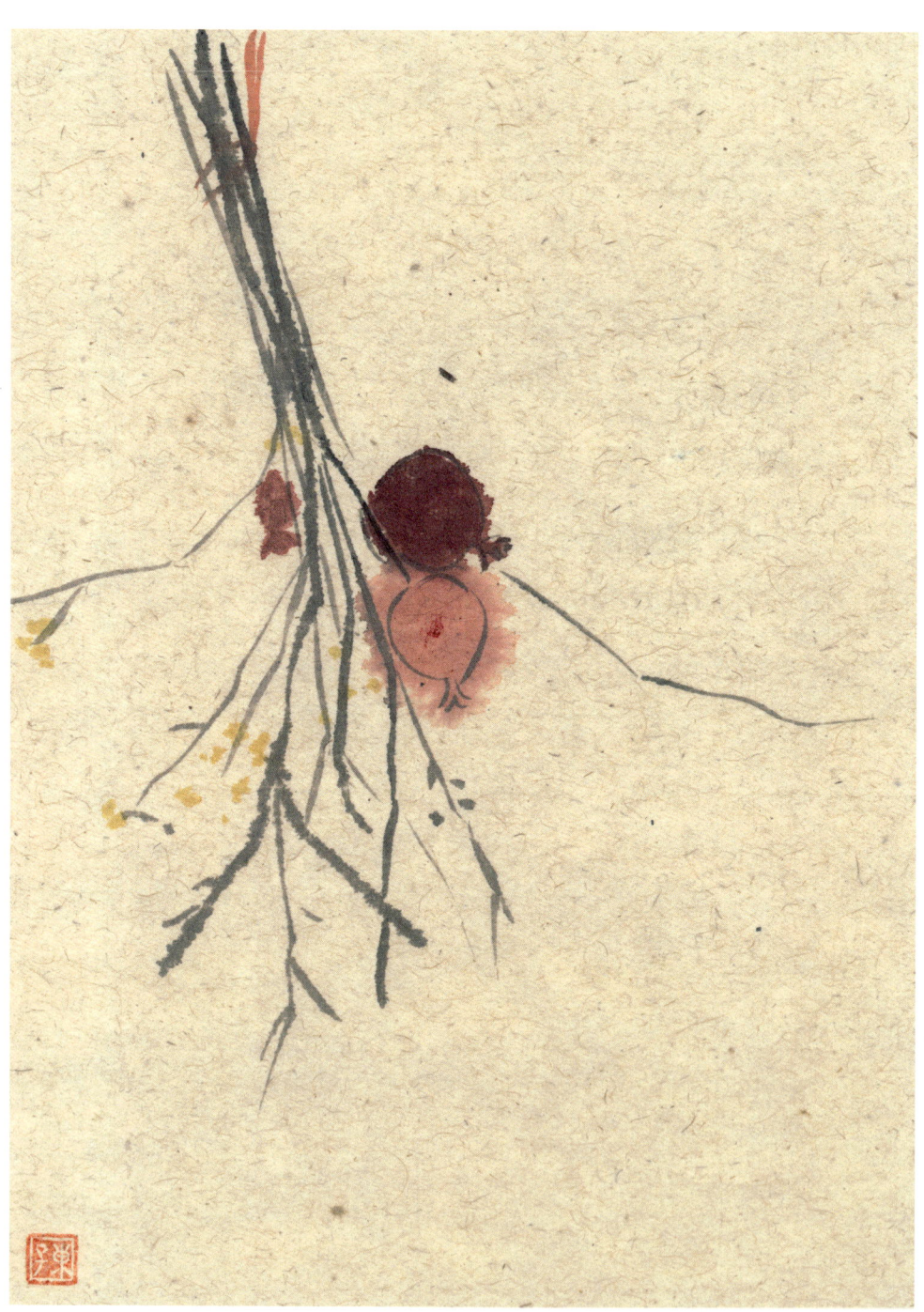

我们只是忘了这里
是早已往返多次的旅游圣地
*We only forgot here*
*Used to be in our holy dreams*
*We had come and go all the time.*

活下来

默默表达

*Express oneself in silence*

*as long as survival.*

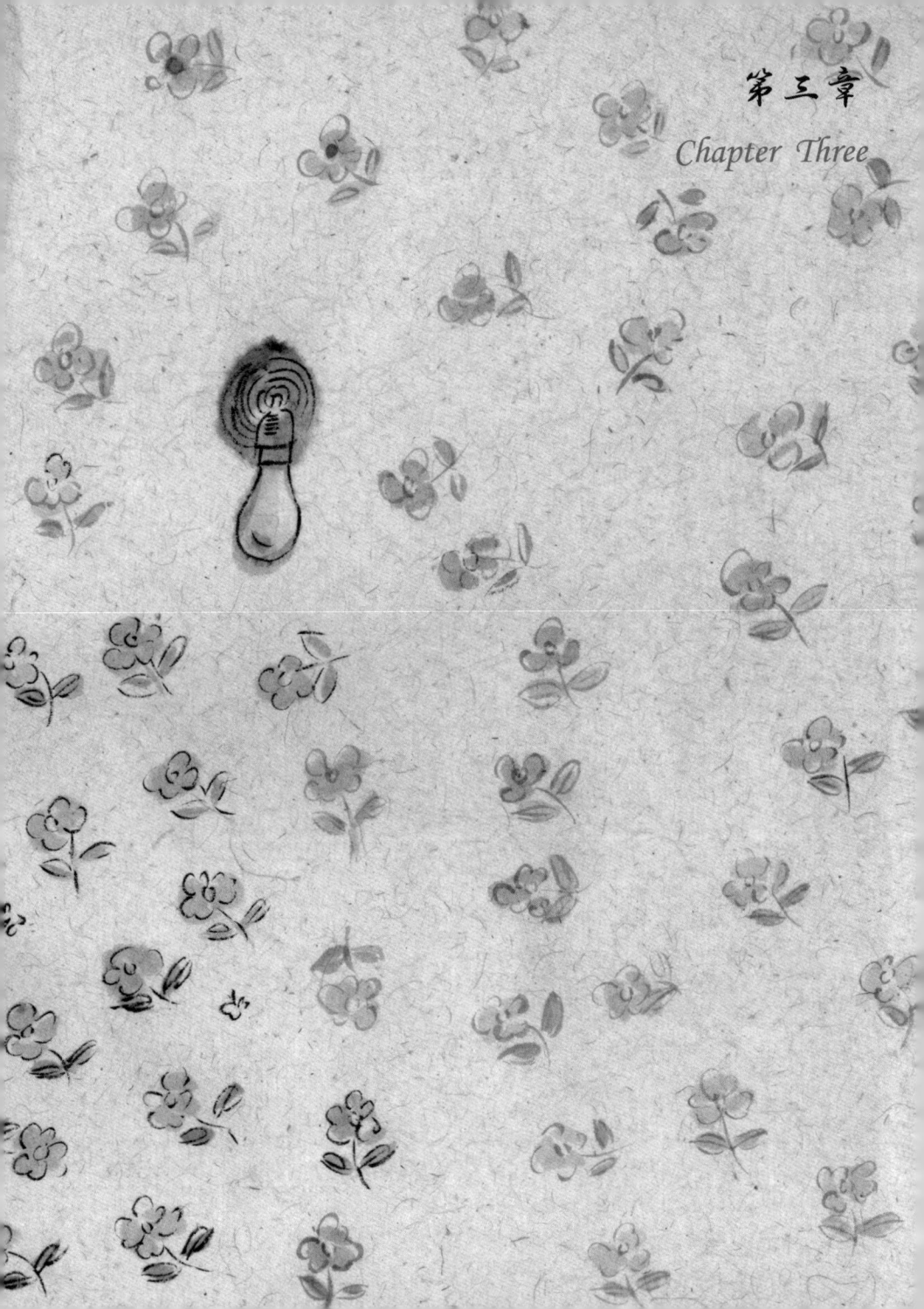

人类就是由思想的纤维构成的蛋体

我们都是会走路的蛋

*Human beings were all sort of eggs made from*

*thinking fibers*

*We are all walking eggs.*

灵魂和肉身是有区别的

一个是司机

另一个是拖拉机

*It is still a lots difference in between soul and flesh*

*one is the driver, and the other is the tractor.*

阳光里变软，我是一粒掉在地上的糖，等着蚂蚁排成队蛇。

70

我是一粒掉在地上的糖
等着蚂蚁排成队来吃

*I was a sugar off the ground,*

*lined up waiting for the ants to eat.*

我会想你从黑夜走来
带着一根有味道的烟

*I am thinking of you*
*and you walk out from darkness, with smell of*
*cigarette.*

遇一好男人

为我编篮子去采花顺道买菜回家

*In meeting a good man*

*He will make basket for me*

*We pick up flowers and buy vegetable on the*

*way home.*

如何度过此生最后一天

*How to spend the last day in our life?*

## 放下的方式很多

*There are many ways to let it be.*

平心静气

*Peace & Tranquility.*

我是你在海边遇到的一只鸟

*I am a bird, you met by the beach.*

# 随照片去做一回苏州菜市场的卖菜女

*Follow the photo, to be a vegetable selling lady, at vegetable market in Suzhou.*

**李玉祥拍的照片**　*Photos taken by Yuxiang Li*

人生就是养的一仙人掌

经常被扎到

*Life is a cactus*

*we often got hurt.*

与你相约八十岁

风和日丽

*Made a promise*

*When we turn to 80 years old*

*I will meet you in a sunny day.*

最难得

人间相聚喜

*What a celebration*

*When families and friends are getting together.*

我看着你站在春天里一直在笑

*I watched you standing in the Spring, have been smiling.*

我不下棋只听风雨

*I will not play chess*
*only listen rain and wind.*

等雪

*Wait for snow.*

源源推松摇雪图

*Painting of Yuanyuan pushing pine tree and shake for its snow.*

第四章

*Chapter Four*

我的灵魂啊
请你每天早晨叫醒我
带领我走向正确的方向

*Oh my soul,*

*Please wake me up every morning and lead me in*

*the right direction.*

# 去见黄昏

*Go to see the Dusk.*

104

黄昏
是让人释然的时刻
今天工作结束
早下班八分钟

*Dusk, is relieved moment,*

*Finished today's work earlier*

*and get off work 8 minutes earlier.*

心如止水

落满线条

*My heart as calm as still water*

*however  be filled with dropped lines.*

当我真正面对独立的时候

寂寞就不存在了

*When I truly face my independence*

*There is no more loneliness.*

天天活在生死边缘
去采希望之花

*Every day I live at the edge of life and death,*
*and to pick up flowers of hope.*

# 佛前喝杯茶

*Drink some tea in front of Buddha.*

桥影提壶图 壶里装着酒 茶酒不分
*Painting of Qiao Ying lifts the teapot. There is wine in the teapot, no difference for either tea or wine.*

男人的心胸
要装得下一个女孩的走动

*A man's open heart*
*would afford his girl's freedom.*

束子

15. 11. 3.

遇一好茶
温暖一下午
遇一好人
温暖一生

*When encounter good tea,*

*warm one afternoon*

*When encounter a good person,*

*warm your whole life.*

用尾巴变一个小岛
看着你在上面光着脚
*Use the tail to become an island*
*You on the island , barefoot.*

原生家庭会影响性格和情绪

但不能占据你的一生

王小方语

*Native families would affect the character and*

*emotions,*

*but cannot occupy your whole life.*

*Quotations by Xiaofang Wang*

灵魂和灵魂相遇
是为了互相补氧

*Soul meets with soul*

*For providing oxygen to each other.*

我的童年一直回响着爸爸爱唱的王二小

放牛歌

牛儿还在山坡吃草

放牛的却不知道哪儿去了

*The song my daddy loves*

*Always echoing near my ear:*

*Cows are eating the grass over the hill*

*However we did not see where is the cowboy*

作词：方冰

*Lyrics by Bing FANG*

# 从虚空里来到无为中去

*From the Emptiness*
*to enter Nothingness.*

一切都是枝头的鸟
在作短暂的停留

*Everything is bird on branches*
*Stop for a short stay.*

天黑了
纸里亮着灯
看书如同给自己打着把伞
躲过人间

*It is dark*

*Light from the paper*

*Reading just like lift umbrella for your heart*

*Escape from the crowd.*

孙可老师插花图
*Painting of Teacher SUN Ke, in preparing for flower-arrangement.*

我们是鸟喂大的人类

*We are the human being fed by birds.*

带翅膀的生命

*Life with wings.*

一滴墨能干好多事呢

*We would do many things*
*Even with one drop of Ink.*

我谨小慎微地站在原地
保持着平庸的样子

*I stand still cautiously*

*maintaining mediocre look.*

找个时间
和光面面相觑

*Find time*
*let us looked at each other in the light.*

人生没有下一件事只有这一件事

*In our lives*
*there is no NEXT thing*
*there is only ONE thing.*

**南京先锋书店老板在泡茶**　*Owner of Nanjing Pioneer bookstore is making tea*

我来人间朝圣

遇到的都是佛

*I come to the human life for my pilgrimage*

*journey*

*Everyone I met is Buddha.*

做完家务思考一下哲学问题

*After house work*

*it is my time to think about philosophy.*

风在天上逃课
我们是睁着眼做梦的族类

*Wind escape from his class in the heaven,*
*We are the kind of day dreaming with open eyes.*

所有美好的东西都是全人类乃至整个生命界共同拥有的财富

*All the good things,*

*are the commonly owned wealth*

*by all humankind and the whole community of*

*life.*

我们的终极课题是如何证明自我的不存在

*The ultimate issue is how we prove that there is no self.*

154

一切都在发生
我们仅仅是宇宙的委托人
*Everything is happening*
*We are only the principal of the Universe.*

我是自由的

追风者

156

我们是自由的追随者

*We are the follower of freedom.*

忘记

是一件多么幸福的事

*What a happy thing to learn how to forget.*

在哪都能遇见猫

*Everywhere, we would meet cats.*

每个人都是一条镜子
彼此照见互相提醒

*Everyone is a mirror*
*We see each other in the mirrors*
*and keep reminding each other.*

时间是用来观察和欣赏生命的成长

*Time is used to observe the growth
and appreciate of life.*

信心清静

即生实相

*Tranquility and confidence,*

*Reflect the true reality in the world.*

早

灵魂相遇没有时间地点

*Good morning*

*there is no time and place,*

*when souls meet each other.*

晚安

写句诗

总结一下自己的人生

*Good night*

*Let us summarize our life by writing a poem.*

愿此刻圆满无缺

*In this very moment*
*Everything is perfect.*

我的大王

王善元

一九三七年

阴历二月二十日

生于

黑龙江　拉哈站

属牛

*My King Shanyun Wang*

*Born on the 20th day of February 1937 by Chinese*

*lunar calendar*

*Born at La Ha Zhan of Heilongjiang province, China.*

*He borned in the year of the Ox in the chinese zodiac.*

人生也是借来的

*Our lives are all borrowed.*

我是我自己
我不是你

瑞贝卡语

*I am myself,*

*I am not you.*

*Rebekka Stocker*

总有一刻被袭击
总有一时相依
总有一天彼此忘记
总有一生在一起

*There is always a moment of being attacked,*

*There is always a moment we are dependent to*
*each other,*

*There is always one day that we forget each other,*

*There is always a life we would be together.*

# 躲在树后的孩子

*Kid behinds the tree.*

# 期待一个紫色的春天

*Wish the Spring full of purple color.*

资国禅茶院内的老丁香开花了

*Old lilac tree is blossoming, at the Zi Guo Tea & Zen garden.*

等你长大的时候

*When you grow up.*

## 大王今天心情很好啊

*The King is so happy today!*

亲爱的太阳

我不知道怎么为你祈祷

*Dear Sun,*

*I don't know how to pray for you.*

存无守有

*Keep the Nothingness.*

194

壶静聚暖

人静凝神

*When the teapot is quiet, it gathers heat.*

*When ones quiet, he gathers concentration.*

# 自由是我的王冠

*Freedom is my Crown*

老眼说话，
乐子制图
2015.1.2

198

不是你的菜

别去揭锅盖

眠公语

*If this is not your dish*

*Do not expose the lid.*

*Quotations by Mian Gong*

因为你是树

所以我想抚摸

*Because you are the tree*

*Thereafter I touch.*

给天空一朵云
给自己一个礼物叫孤独

*Give a cloud to the sky*
*Give yourself a gift name as loneliness.*

204

自在

自己不在了

*Being comfortable and free*

*then the self will disappear.*

206

感受力回归源头

*Sensibility returns to its original source.*

体有金光

视而不见

*There are golden light inside the body*

*and to pay no heed.*

去心里爬梯

*Climb the ladder in one's heart.*

# 相遇在冬天的精灵

*Elves met in the Winter.*

风动水不愁

*Wind stop*

*Water will not be sorrow.*

活着不难
好玩不易

*It is not very difficult to be alive*

*however it is never that easy to always have fun*

*in one's life.*

阳光明亮

有人在捞鱼虫

生活有时比冬天的河水还凉

*On the sunny day*

*I see someone is picking up food for fish from the river*

*Sometime life can be even freezing than the water in Winter.*

人生是看别人的剧本
写自己的台词

*Life is first look at someone else's script*
*then writes their own lines.*

# 松下成就图

*Reach accomplishment under pine tree.*

孤独像根

让人活得很深

*Loneliness is like root*

*make us live even more deeply*

每棵树长成什么样

都有自己的想法

*Each tree would has its own idea*

*about how she would like to grow up*

我们什么缘份
我怎么有点想你

*We have predestined infinity*
*how come I start to think of you.*

你的孤独是我的家

*Your lonely is my home.*

万事皆有阴阳

每天自有因果

*All things are Yin and Yang*

*Everyday there is reason and result.*

生活清贫不是你的错

生活脏乱是你的错

　　　　　　　一丹语

*It's not your fault to be poor*

*however it is your fault to be messy*

　　　　　　　*Quotations by Yi Dan*

236

玩是最高境界

*Play is the highest realm.*

瑞贝卡画的树　*Rebekka stocker*

如果你是魔鬼
我就是你的心
如果你是天使
我就是你的翅膀

*If you are a devil*

*I am your heart*

*If you are an angel*

*I am your wing*

心越画越大

人越画越小

*Heart will become bigger and bigger with the creation of painting.*

*People will become smaller and smaller through the creation of painting.*

人不能闲着也不能吃得太饱

*Man can not be too idle or eat too much.*

好久没回大自然

*Have not been back to Nature for so long.*

# 后 记
## Epilogue

愿我从精神花园里采摘更多的花朵带往人间

*I am more than happy to pick flowers*
*from my spiritual garden.*

（京）新登字 083 号

图书在版编目（CIP）数据

大王今天很高兴:东子哲学绘本/东子著. —北京:中国青年
出版社,2015.4
ISBN 978-7-5153-3244-4

Ⅰ.①大…　Ⅱ.①东…　Ⅲ.①人生哲学–通俗读物　Ⅳ.①B821–49

中国版本图书馆 CIP 数据核字(2015)第 072491 号

责任编辑:申永霞
排版设计: JYT-Printing

\*

中国青年出版社 出版 发行
社址:北京东四十二条 21 号　邮政编码:100708
网址:www.cyp.com.cn
编辑部电话:(010)57350501　门市部电话:(010)57350370
北京顺诚彩色印刷有限公司印刷　新华书店经销

\*

880×1230　1/32　8 印张
2015 年 4 月北京第 1 版　2015 年 8 月北京第 2 次印刷
印数:3001–6000 册　定价:38.00 元
本图书如有印装质量问题,请凭购书发票与质检部联系调换
联系电话:(010)57350337